Ebony and The Four Frogs

Masters Addition

By Donald E. Thomas

"La, La, La, La, La, Laaaaaaaaa" the melodies rang loud in the kitchen as Ebony and her dad sing their favorite song together. The kitchen is buzzing, as excitement fills the room from top to bottom. Dad, is at the sink washing the dishes from the delicious meal they just ate. Suddenly, Ebony's big grin turned into a frown as she takes a closer look at her assignment. "Is everything okay Pumpkin?" asks Dad with concern. Ebony takes a deep breath and says, "I can't do addition." Dad speaks in a kind voice. "I'll help you in a few minutes." Their dog Chico, notices that his buddy is not very happy. He is not a math specialist, but he knows Ebony is in need of help.

Dad quickly jumps in to help Ebony with her math problems. He smiles as he tries his best to lift her spirits. Dad says, "Okay, what is 4+3?" Ebony is confident as she sits up tall and says, "it's 13. Yep, 4+3= 13".
Dad said, "No, honey that's not right." Chico tilts his head sideways as he knows 13 is the wrong answer.

Dad is clearly frustrated. He can't understand why she isn't comprehending the process of addition. Dad says in a stern voice, "Sweetheart, we have been working on this for quite some time now." Ebony is trying her hardest but just can't seem to grab a hold on addition. Ebony begins to speak in a shaky voice, "The boys at school said girls are not good at math." Chico covers one of his ears with his furry paw, as though he can't believe what he just heard.

After her bath, Ebony climbed into bed. The moonlight streamed through the window. Dad steps into the doorway and softly says, "Goodnight, Pumpkin." Goodnight, Dad, Ebony mumbled, rolling toward the wall.

Ebony tossed and turned in bed as she remembered the time when she got a math problem wrong in class, and the boys sitting behind her laughed. One of them whispered, "Girls can't do math." A look of displeasure crossed Ebony's usual happy face. Now Language arts is her favorite subject and Social Studies is her thing. So, for a quick second she thought, "um maybe the boys are right. Maybe girls can't do math." However, Ebony is a competitive student and she is determined to prove those boys wrong.

Ebony and Chico finally fall fast asleep. The room is still when suddenly, something magical begins happening in her bedroom. The butterflies on the wall seem to have come alive. They're actually flying! Ebony's basketball starts to spin by itself, then stops abruptly. There was a flash of light, then SHA-ZING! shadows of frogs begin dancing on the ceiling. Amongst the magic swirling in Ebony's room, you could hear the sound of a squeaking door opening. The sound grew louder and louder. There is a secret door inside of her room! Despite the excitement stirring, Ebony, Dad and Chico are still sound asleep.

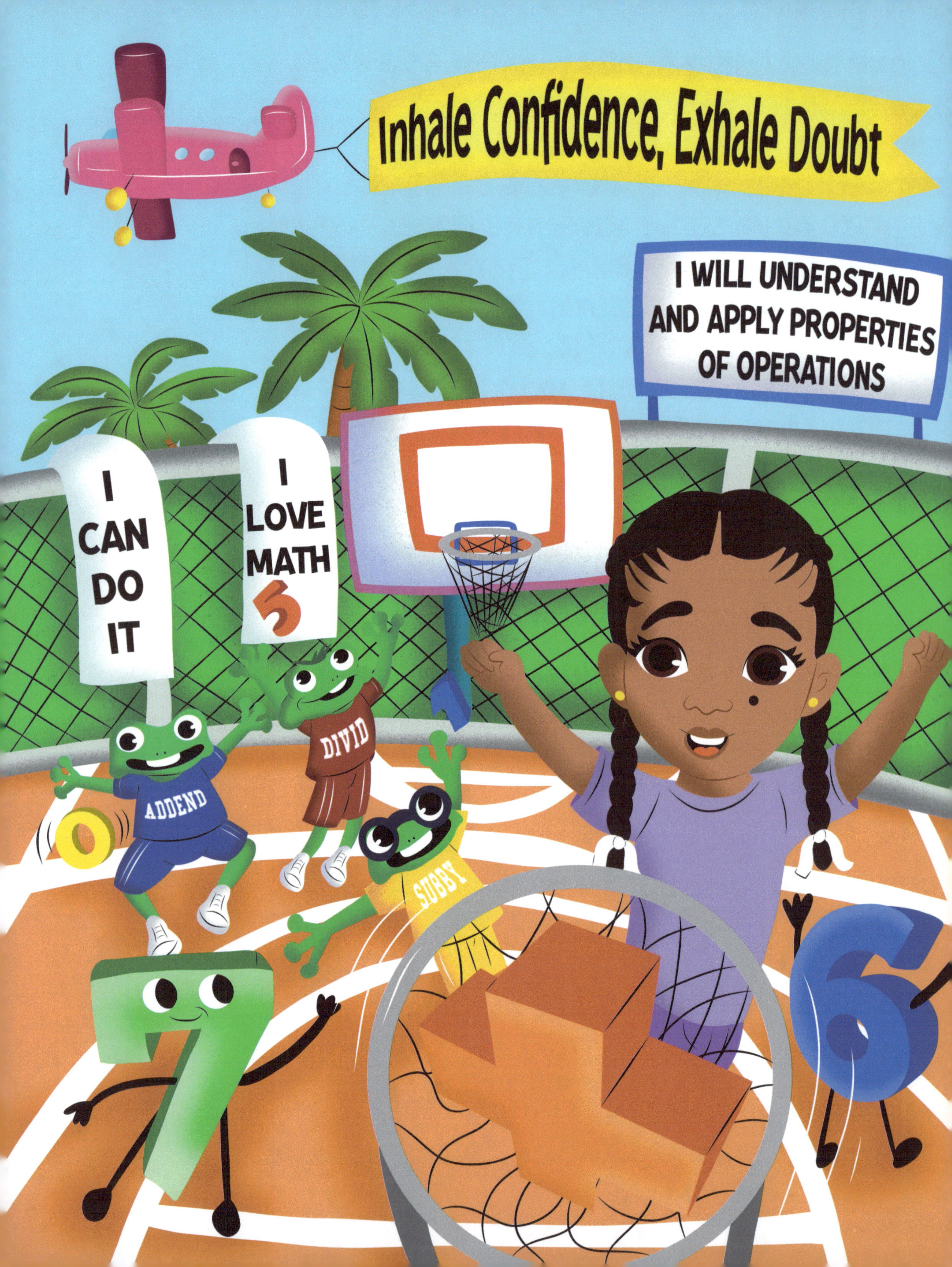
Inhale Confidence, Exhale Doubt
I WILL UNDERSTAND AND APPLY PROPERTIES OF OPERATIONS
I CAN DO IT
I LOVE MATH
ADDEND
DIVID
SUBBY

You can hear music playing as Ebony tumbles into a magical math land. Four frogs are jumping and dancing around shouting, "Ebony, you can do it." There are talking strawberries with faces marching in line. Each one of the frogs jumped extremely high and introduced themselves. There is Divid, Multi, Suby and Addend. Addend says "I will be helping you to learn addition." Ebony's attention moves to a pink airplane displaying a banner. Ebony begins to read it out loud, "Inhale Confidence, Exhale Doubt". She takes a deep breath and feels confident immediately.

The air is warm and the sky is clear, Addend said to Ebony "take a deep breath, inhale Confidence, and Exhale Doubt". Now Addend knows that basketball is her favorite thing to do. So, he begins, "If you are playing in a game and you have 8 points and you score another basket how many points do you have now?" Ebony takes a deep breath and shouts "10". "That's it, this strategy is called, counting on. We will use this skill when doing addition" Addend proclaims.

ADDEND

In Magical Math Land there is a basketball court which is the key to unlock Ebony's love for math. Addend shouted out "7+3=?" Ebony replied "10". This went on for quite a while. Ebony is building her confidence while embracing a new found love for math.

Ebony was slam-dunking facts now! "What is 4+X=12" Addend asked with a sly grin. "8" Ebony shouted without hesitation. She figured it out using mental math.

ADDEND

As Ebony and Addend walked along the math road, there are signs to remind her to "Count On" and Memorize the Facts!" The strawberries that Addend used to visualize sums are having a tug of war contest. Ebony smiled at the strawberries and said, "3+3=6." Suddenly, the squeak of a door echoed in the air, and with a whoosh, Ebony was back in her bed.

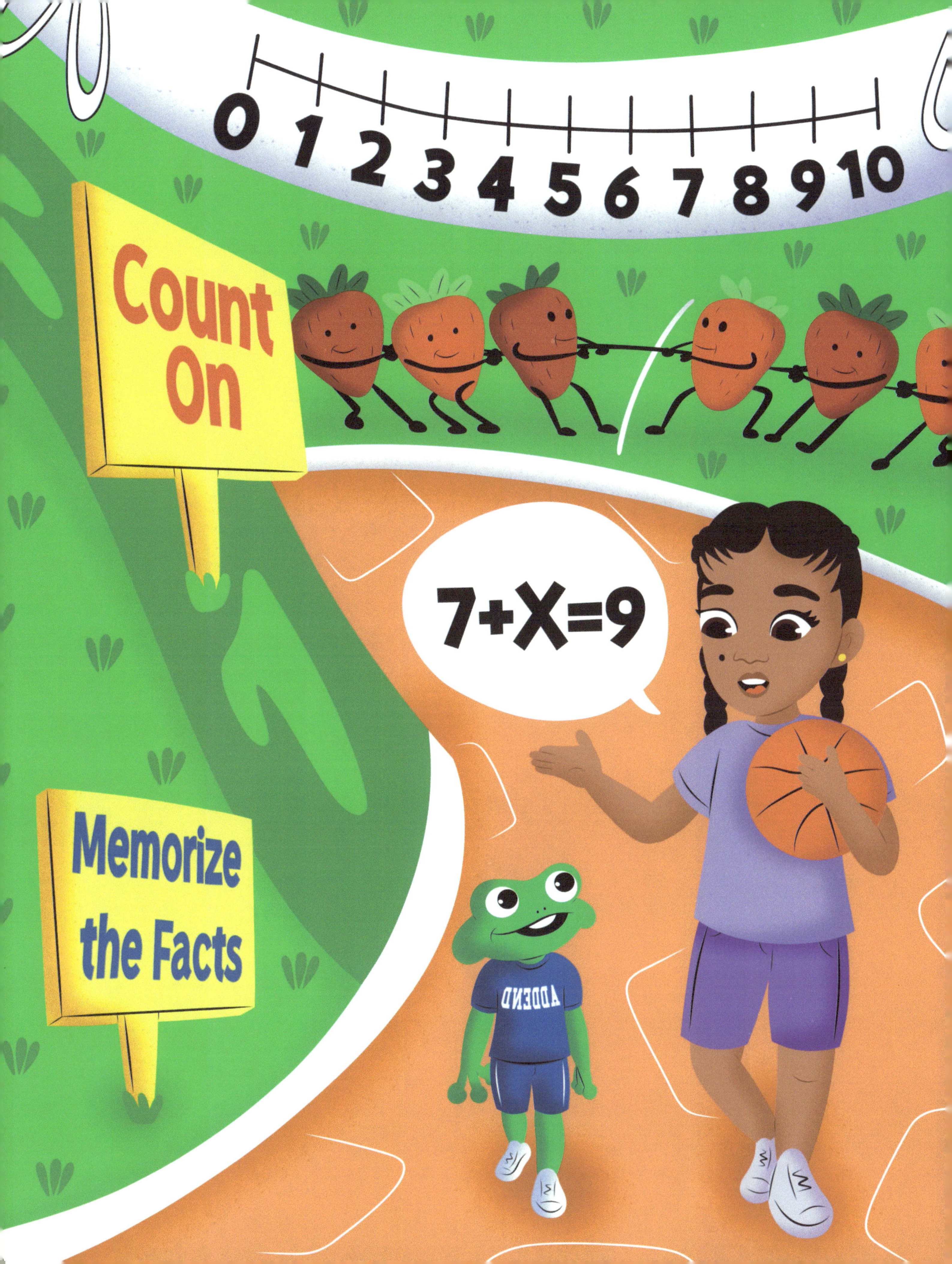

0 1 2 3 4 5 6 7 8 9 10
Count On
Memorize the Facts
7+X=9
ADDEND

5+4=9
8+4=12

The morning sun warmed Ebony's face as she woke up, shouting math facts with a grin. Chico danced around the room, his tail wagging with joy.

12
1
2
3
4
5
6
7
8
9
10
11

At breakfast, Ebony used strawberries to solve sums. Dad chuckled, "Why didn't I think of that?" He couldn't explain what happened overnight, but he knew something magical had transformed his daughter.

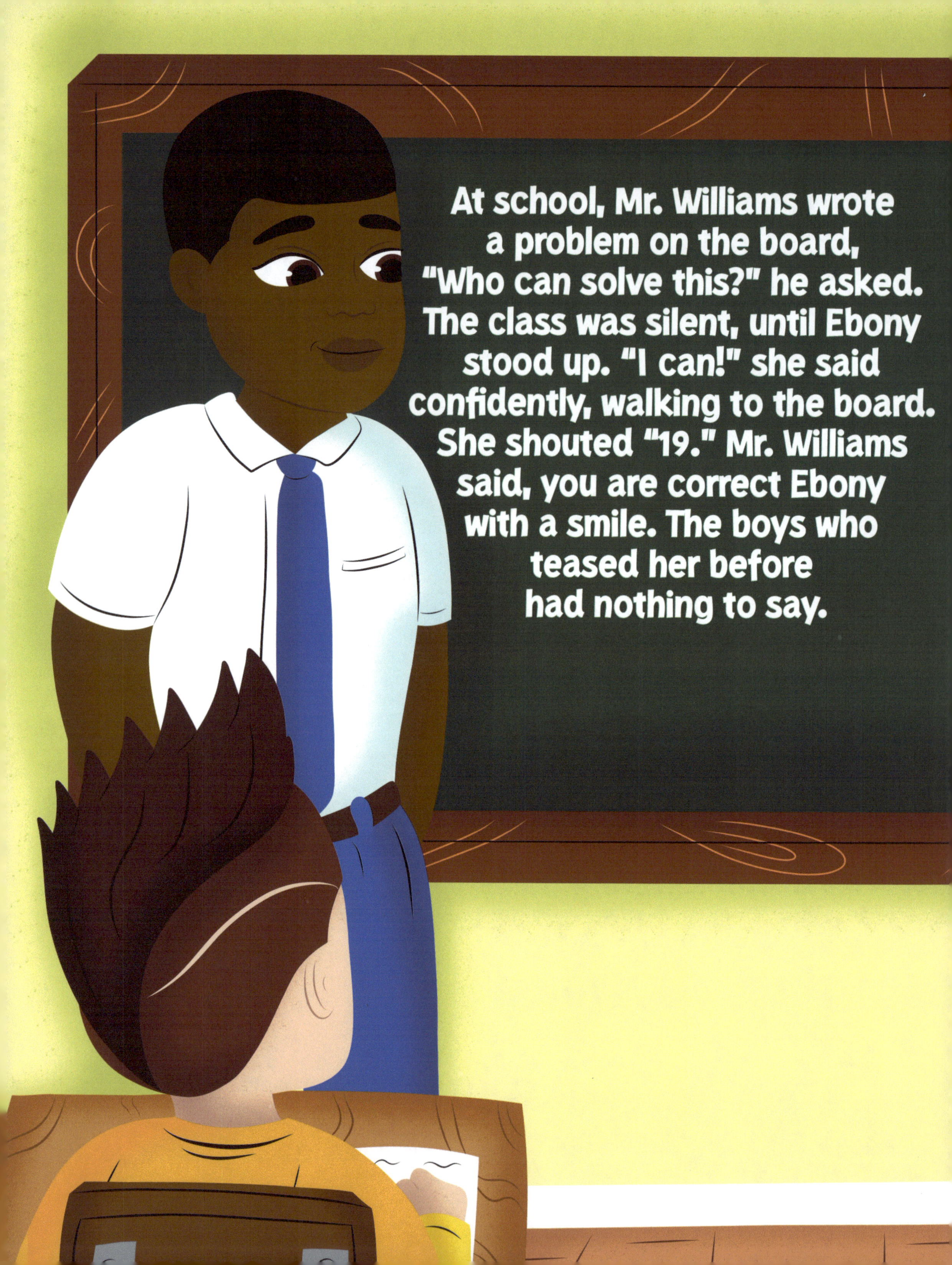

At school, Mr. Williams wrote a problem on the board, "Who can solve this?" he asked. The class was silent, until Ebony stood up. "I can!" she said confidently, walking to the board. She shouted "19." Mr. Williams said, you are correct Ebony with a smile. The boys who teased her before had nothing to say.

If Jordan had 14 acorns
and he found 5 more.
How many does he
have in all?
15 16 19 17 18

During recess Ebony stayed in class to practice math. She dozed off at her desk and tumbled back into Magical Math Land, where she and Addend flew through the air shouting, "I can do math!" Ebony had slam-dunked addition and left doubt far behind.

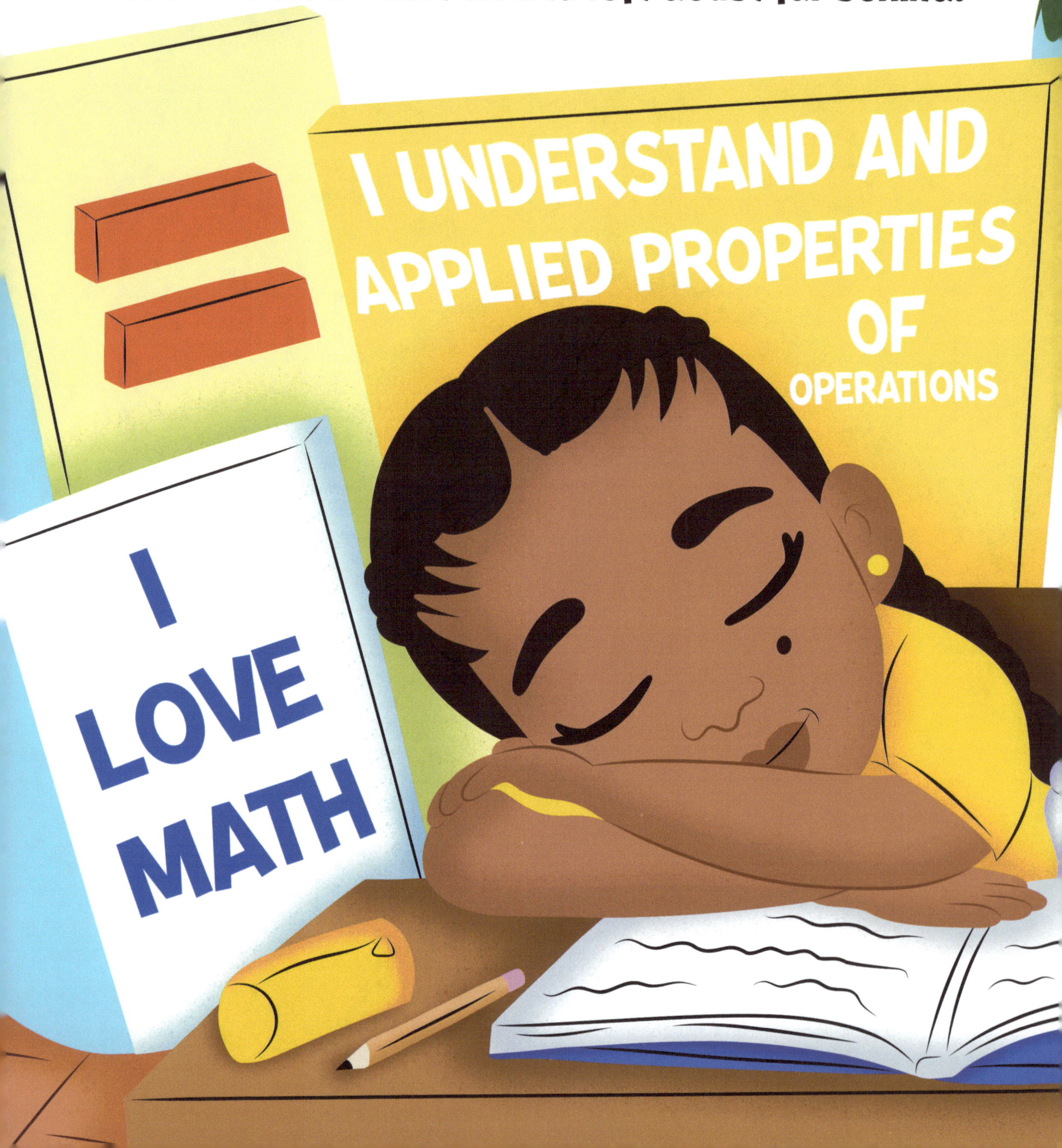

I GOT THIS

For My Daughter Dana

Library of Congress Control Number:
9798991698504

ISBN: 979-8-9916985-0-4 (paperback)
ISBN: 979-8-9916985-1-1 (hardcover)

www.ingramcontent.com/pod-product-compliance
Lightning Source LLC
LaVergne TN
LVHW071248210926
848486LV00008B/367